When The Light Came On Regarding Our Historical Past

This short publication is called Defining Human; it is what happens

when the heavens send you down an omen or incredible luck.

The Vision Seen

Kanata, Ottoman Linguistic Analysis

Animal To Hominid Traits

Radicalization of Language

Social Identity & Culture

The First 9 Females (singular-related family DNA) Clans Who Crossed

Who was the first Man

The Conveyer-belt theory in their perspective Landmasses

Radicalization of Language

Defining Human

Progressive environmentalism

Defining a Constant Of Habitat vs. Settlement

The first half of this publication is called the section on Pictionary Descriptive, no different then the sounds to visual attachment, seen by our ancestry one million years ago with the one exception, now we have technology and full dialogues.

It is done with a twist of humor, the primitive versus new. The second part is a **preview of my anthropology notes** on the component of humble human beginnings. This would be in defining human traits research, sources and the development of race.

As simple as is it, the lack of historical acknowledgement missing is enormous. The concept that we are one single family was never fully understood other than the possible suggestion from a few old Native books that I have seen.

The ideology that we have to live in the breathing walls known as the super organism called earth also comes to us also from Native philosophy.

This report was generated to create an impact for change! To start the dynamics of what is called progressive environmentalism.

Once upon a time, a relatively low-key person started hobby research, she got jailed.

They told her this <mark>was way too much research for one person</mark> and it would upset everyone including, our Canadian Natives.

...... Pardon?

The richest families in Canada secretly lobbied the government due to the content of

Ottoman; then they sold it to the United States and Turkey.

Give it back, especially Kanata was my response!

Now under Ottoman, I was blocked from generating revenue and my budget is tight

But we are still governed by Law.

This is for Landscape Betterment Projects (LBP) and not your pockets.

But we want a Native girl, they said.

They also said I was too Turkish and because of my children

I should "<u>get it</u>" why this research can't be given to me.

Furthermore, an independent thesis' does not count; they continued.

a University BA was just not sufficient.

You have to have PH. D in the standard technical sense?

I responded, after Obama are you still at the race-game card,

and besides what is your definition of too Turkish?

I AM CANADIAN

Also I am a humanist, and free, I responded. This is the 21st Century.

Ottoman brought out hundreds of questions on race and language.

What? This has now turned into a Dr. Martin Luther King tragic event?

Define Human I said?

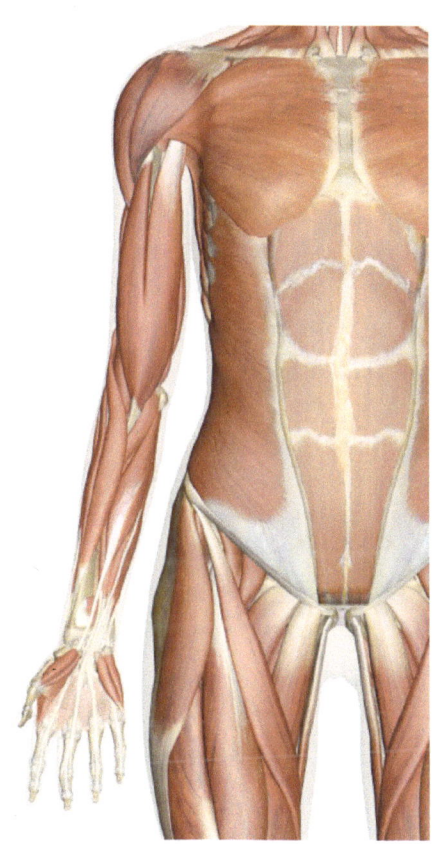

Ok for the thousand time, let's start at the beginning.

I told them: My dad trained me on the Ottoman Empire.

No, I am absolutely not political or working as some secret agent for Turkey.

Are you actually kidding me?

Let's get realistic the world of James Bond is also impacted by socio-economics.

I was technically orphaned at 10 and grew up predominantly with the black

community my entire young adult life. For years, I was protected under

The Metro Toronto Housing Authority.

I am studying human social behaviors and betterment pilot projects. Why humans do

certain catalyst to an array of violence in our surroundings. I insist this is academic.

And what about the women who did not have a degree

and studied monkeys all her life.

I can sincerely relate with her.

How valuable, she was for animal conservation?

And yes during that period, I was under distress but will it make you feel better

that the discovery of Kanata Means Wings (cover), was coincidentally discovered on

Canada's birthday July 1st, 2013.

Not just for marketing.

We can use all of this for pushing green Native philosophy.

Remember Mr. Atheist, I told them:

What did I pray for in church?

I prayed to the skies to be better than those before me.

You work, you get, is exactly what happened!

I am the womb!

A hardcore environmentalist for years.

I work hard and I am corporately trained!

We need changes look at all of this as a blessing.

I told Canada

Release me immediately and unblock all my information.

How dare you attempt to dissect it.

I just saw the naughty things that historical politics (via academic manipulation) for

the last two thousand years have done.

No this is not mental illness.

It was just my lucky visit in dark times to a house of God or a one in

a trillion chance because of my multifaceted background.

You may choose what you wish to believe.

I told the Americans

You know what you did and yes

I got upset and changed the Middle East, but academics and

politics have to be separated.

Please forgive me for the index!

This is a Canadian show now.

§

I continued, please don't turn my academic work into a political brawl, and

what about the Armenian Genocide? That was 100 years ago?

§

This project belongs to the making of a better world

Wink the money of social betterment I told them.

I had to say money to the Americans!

Furthermore, find out who ==spent a lot to impersonate me.==

I am extremely private and classy. These big players are mortifying.

What do you mean Russia and China want to nuke me now?

Ottoman is also a hundred years old.

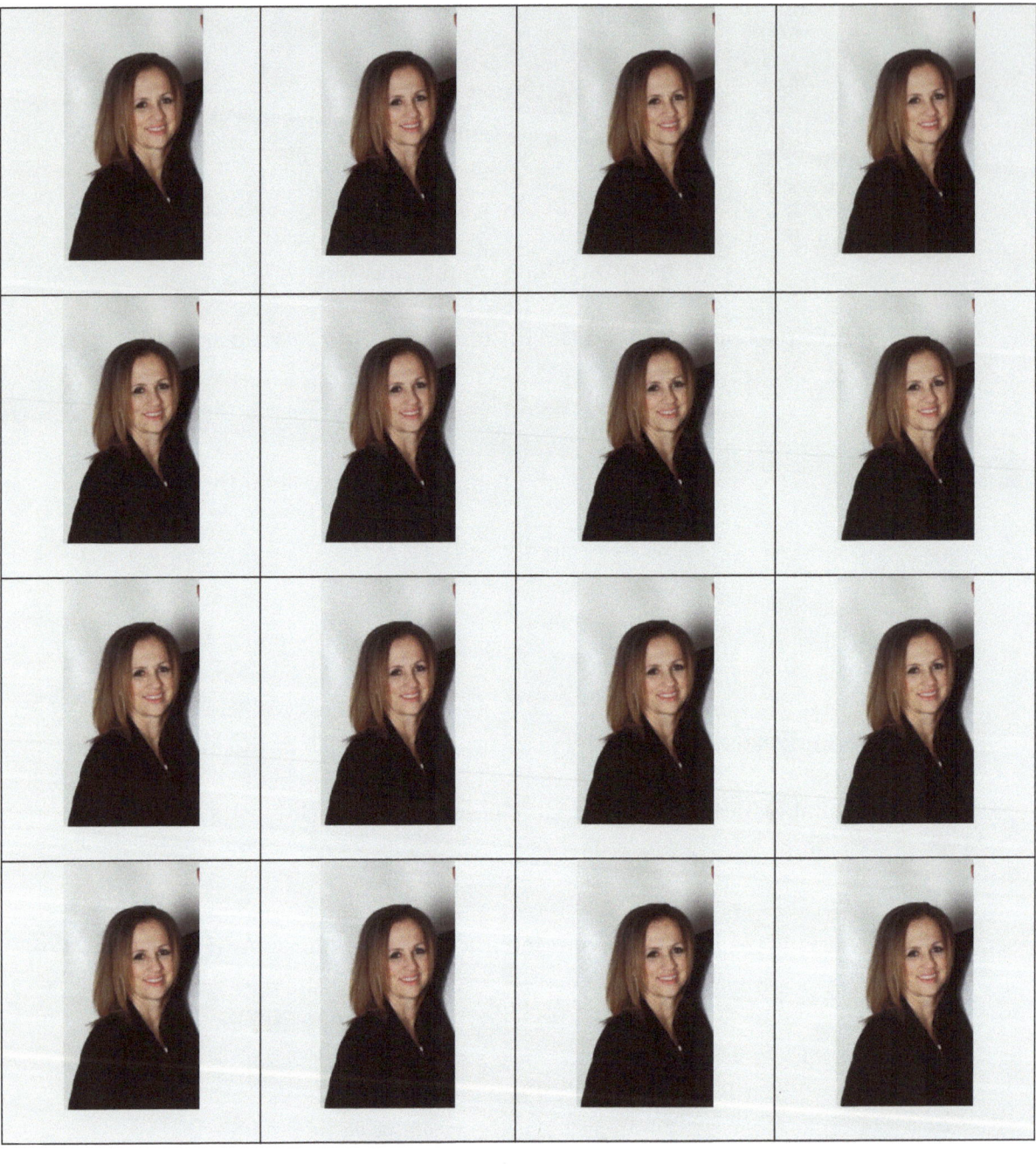

I told Israel

You may think Yerushalayim, (Jerusalem's ancient name) is Hebrew

but it has a shamanic connotation that goes back possibly 100,000 years.

And it is linked to both Kanata and Ottoman.

Read what I have written carefully.

It is a beautiful cultural discovery; please can you just get over it.

Pre-review, Royal Distress I Felt.

Ok it is a word transposed, do you feel better I told them?

My children, my background, linguistic knowledge on shamanic Turkic countries, to that of words globally found, made me see that

The following picture below are actually

<u>two land masses </u>with the breaking of the Bering Strait:

1   2

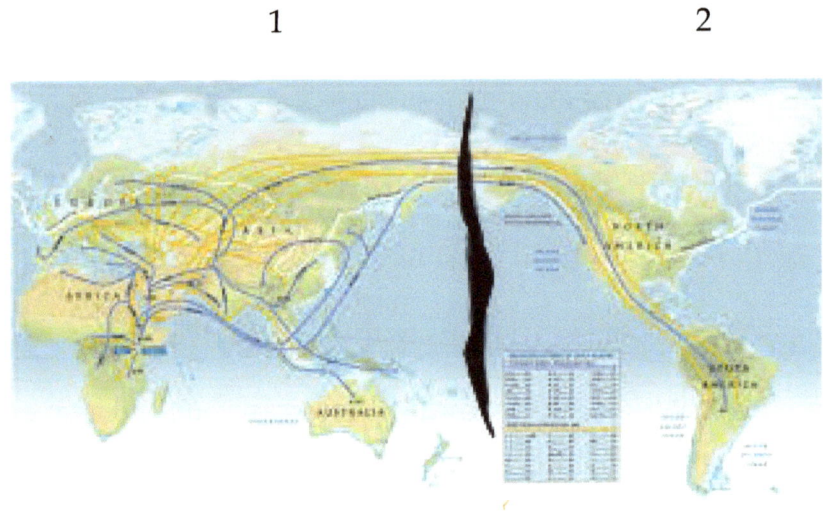

We are getting there!

My children helped so much they looked beautifully shaman.

I am mentally processing the map above.

Natives were technically on new undamaged lands at less than 50,000-year.

Black skin is climatic.

But there are so many varieties of human how could the prototype like my children be

the only first people of earth?

Each landmass is our focus now. There is a dominance on Landmass two?

This is getting worst.

Look at land again, my brain tells me...

I now start giggling…

Hey Geneva, you may be white but your ancestors were not!

More jail time

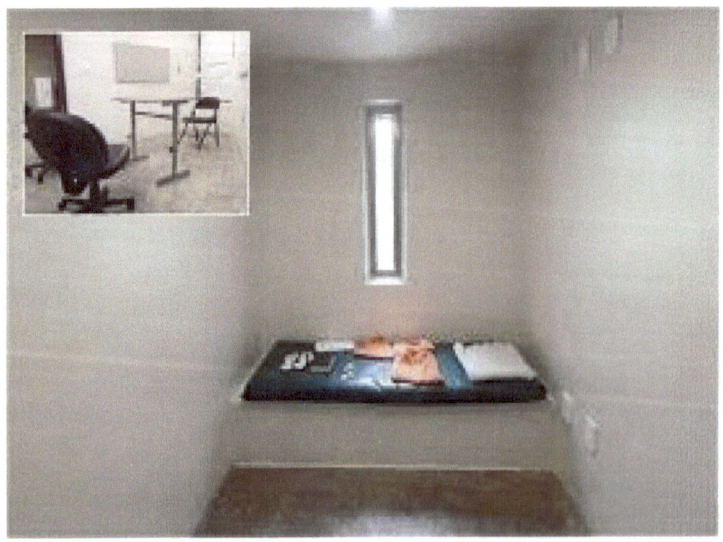

Be smart I thought.

William Shakespeare might have rolled in his grave at the speed 300 poems on our

global shamanic history were composed.

My work was flipped to all the universities, pronto.

After they read the notes

they passed right out.

Prof are you ok?

Child they said are you being irresponsible, how do you know?

It is called linguistic anthropology, and the relooking at the link of languages globally.

The key is my hobby research of covert operations of Military, heritage knowledge of

Anatolia and several Island phonetics.

Ottoman and Kanata just sealed the deal.

All of you should thank me for being so damn fair.

It is called the 50-year Mega project for phonetics to DNA tracking.

This is our global heritage.

By the way, now that I am out and this is in the courts.

Can you get rid of my Pocahontas-Barbie actress replacement I said?

I am quite offended.

What happened to the Codes of Law that governs us?

No this is not acceptable.

I am threatening to take it to the highest Human Rights Courts in Switzerland.

I am highly educated, thank you!

Can you define human and skin, I asked them again?

Part Two

A Quick Overview- Academic portion

The Variances In Our Hominids To Humans

From a single bacterium a series of transformation occurred on earth.

Today some of the oldest constants of habitats and their lifestyles of clan living and social behaviors of intermarrying or inter-mating gives us a good overview of how human evolution started developing.

Was Darwin's Evolutionary chart therefore not fully accurate?

The answer is yes!

There were thousands or millions of monkeys and hominids involved.

His famous chart is now a pyramid with monkeys at the bottom.

Can you visualize the following?

1) Primate Races 2) Saturation 3) Human Traits Development

You will love the second edition!

In the process of being reviewed.

The concept that different monkeys and their traits contributed to our earliest humans

Known as hominids, that pulled in a form of dinasauric genetic variety.

Bacteria evolved to split ==and attach==.

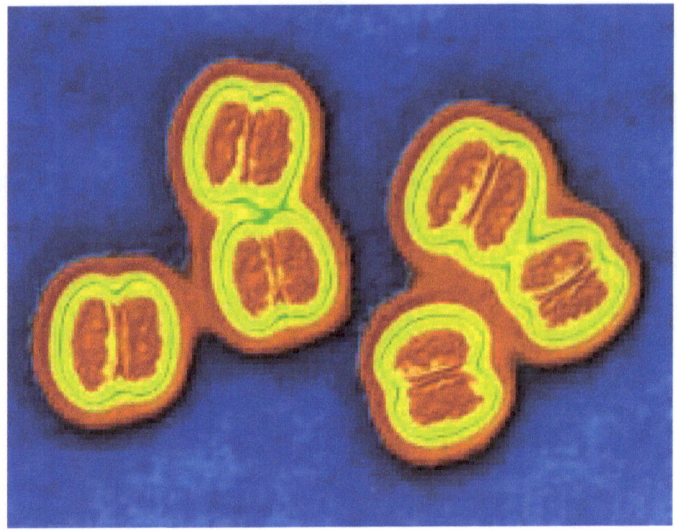

So did our earliest humans and animals, even in social behaviors.

Guess what civil rights 101 presents today

"We are all one family"

Topographically molded.

Sounds silly but before my special time with my friends,

academics said we once came from different sources.

Do you want more?

Was your monkey better than mine?

Humans came to earth in various forms due to animal

traits from clans and saturation.

This happened initially like the bacteria picture above via

mutations from close blood line intermixing.

Called self replication followed by millions of years of social behaviors continuing.

The Ulas family inter-related family came out on all fours is a good example.

A reverse mutation.

A bloodline too close producing what was once normal historically.

Furthermore, from them,

only the strongest survived.

We are building humans from start to finish.

A futuristic finish is needed.

Feel the black earth concept guess what, that same unprotected earth today

Is billions of years of once living ancestry on hard rock.

Webpages: **www.starilkin.com** / www.vancats.com

By Canadian Author

Yildiz Ilkin